DIEGO

THE GALÁPAGOS GIANT TORTOISE

WRITTEN BY

Darcy Pattison

ILLUSTRATED BY

Amanda Zimmerman

Saving a Species from Extinction

Diego, the Galápagos Giant Tortoise: Saving a Species from Extinction
Book 5, Another Extraordinary Animal series
Written by Darcy Pattison
Illustrated by Amanda Zimmerman

Mims House
1309 Broadway
Little Rock, AR 72202
MimsHouseBooks.com

Names: Pattison, Darcy, author. | Zimmerman, Amanda, illustrator.

Title: Diego, the Galápagos giant tortoise : saving a species from extinction / by Darcy Pattison; illustrated by Amanda Zimmerman.
Series: Another Extraordinary Animal
Description: Little Rock, AR: Mims House, 2022. | Summary: An Española Galápagos giant tortoise helps save his species from extinction and then returns home to his island.
Identifiers: LCCN: 2021921657 | ISBN: 9781629441870 (Hardcover) | 9781629441887 (pbk.) | 9781629441894 (ebook) | 9781629441900 (audio)
Subjects: LCSH Galápagos tortoise--Juvenile literature. | Endangered species--Galápagos Islands--Juvenile literature. | Turtles--Juvenile literature. | Natural history--Galápagos Islands--Juvenile literature. | BISAC JUVENILE NONFICTION / Animals / Turtles & Tortoises | JUVENILE NONFICTION / Science & Nature / Environmental Conservation & Protection | JUVENILE NONFICTION / History / Central & South America
Classification: LCC QL666.C584 P38 2022 | DDC 597.92/46--dc23

Sometime in the late 1600s, a pirate ship dropped anchor off the shore of a flat strip of land named Española Island. It is one of the smaller Galápagos Islands, which lie off the coast of Ecuador in South America. The pirates lowered a small boat to the water. Sailors rowed to land, searching for fresh food.

They discovered...

...thousands of giant tortoises!

Maybe as many as 8,000 tortoises.

From 1700 to 1900, ship after ship packed the giant tortoises into their holds. The tortoises could live for up to a year with no food or water, providing fresh meat for sailors to eat on a long journey. Year after year, tortoises were taken until most had disappeared.

In 1905, an expedition from the California Academy of Sciences visited Española. They tramped across the island for three days.

The scientists saw
only
three giant tortoises.

Despite the low numbers, sometime around then, a female Española tortoise laid a clutch of eggs. Months later the hatchlings climbed their way upward and broke through the dirt to the surface.

A dark shadow cut across the nest. A hawk dived and snatched up one of the hatchlings. Another emerged and scrambled under a bush to safety.

Day after day, year after year, that tortoise ate, slept in the shade, and avoided the hawks.

Between 1934 and 1936, the *Velero III*, a California research boat, visited Española Island several times. The scientists on the boat were studying the marine life of the Pacific Ocean. At some point, they probably collected the tortoise, which by then was a young adult.

When the ship returned to California, the tortoise was taken to the San Diego Zoo to join its tortoise herd. He was 20 to 30 years old.

The zoo painted a white number on the tortoise's shell: 21. Day after day, Number 21 looked back at children and families who stared at him. Year after year, he ate, slept in the shade, and lived at the zoo.

In the early 1960s, scientists considered the Española tortoises to be extinct, or nearly extinct. But in August 1963, a conservation officer visited Española and saw 15 goats eating cactus. And in the middle of the goats was a male tortoise!

The scientists wondered if more tortoises had survived. Over the next 10 years from 1964 to 1974, they collected 12 females and two males.

Only 14
Española tortoises
were left
in the world.

The tortoises were taken across the ocean to the Tortoise Breeding Center on Santa Cruz Island, another island in the Galápagos. Fourteen individuals might not be enough to save the species, but scientists decided to try. They started the Española tortoise captive breeding program.

It was difficult. At first, few eggs were laid. Even when there was a nest, few hatchlings survived.

Observations of tortoises in their natural habitat helped the scientists make changes at the Breeding Center. They dug new nesting sites and filled them with a fine soil at least 12 to 15 inches (30 to 40 cm) deep. They changed the tortoises' food to make them healthier and collected the eggs to place them into solar-heated incubators.

The 1970-1971 nesting season was successful.

20 Española tortoises hatched.

The baby tortoises couldn't return to Española Island until they were four to five years old. By then, they would be too large for the hawks to eat and would be able to survive other threats.

Finally, in 1975, the park rangers loaded their motorboats and traveled to Española Island.

17 young tortoises were returned to Española Island.

It was a beginning.

The biggest problem was that there were only two males to father the hatchlings. When they heard of the Española breeding project, the San Diego Zoo invited Galápagos scientists to visit and study Number 21.

Some Galápagos tortoises have a dome-shaped shell, like a ball that has been cut in half. It is perfect for tortoises that eat low-growing plants and grasses. However, Española tortoises are a smaller species with a saddleback shell. Their shells are longer and flatter. The front rises into a big arch over the tortoise's neck. The arch allows the Española tortoises to lift its necks higher to feed on the lower branches of small shrubs or trees.

Scientists decided that Number 21 was an Española tortoise!

Now there were 15 adult tortoises!

In 1977, scientists brought Number 21 to the Tortoise Breeding Center. He was at least 65 years old.

Number 21 was soon renamed Diego, after the zoo where he had lived for over 40 years.

Now he lived with the other 14 Española tortoises, eating, sleeping in the shade, and enjoying the tropical weather of Santa Cruz Island.

Soon, Diego became a father!

When they were old enough, Diego's children were released on Española Island, along with other young tortoises. There, the young tortoises grew up and became mothers and fathers.

In June 1991, the first live hatchlings were found on the island. Diego was a grandfather!

And, for the first time in over 80 years, the tortoise population on the island was growing by itself.

In 2019, scientists counted the tortoises living on Española Island.

2,354 tortoises!

Scientists estimated that Diego was the father of about 900 to 1000 of them. About 518 of the tortoises had hatched on the island.

The Española tortoise would not become extinct!

But there was one more thing to do. Scientists decided that the captive breeding program had been a success. It was time to shut it down. Diego and the other 14 original tortoises had slowly, year after year, rebuilt the population of their species.

It was time to send them home.

On June 15, 2020, at 1:00 a.m., scientists loaded the old tortoises onto a boat for the trip from Santa Cruz to Española. Under a crescent moon, the boat raced across the waves. It arrived at Española at about dawn.

The tortoises were then strapped onto park rangers' backs. The female tortoises were lighter, easy for one person to carry. The three males were heavier so two people took turns carrying each of the male tortoises.

Park rangers had found a place with lots of *Opuntia*, or prickly pear cactus. It was the tortoises' favorite food. It took two trips from the beach to carry all 15 tortoises to the cactus.

When the tortoises were released, they ignored the park rangers and started feeding. Perhaps they were excited to eat plant species they hadn't tasted in years.

Diego had lived at the San Diego Zoo for about 44 years, and at the Tortoise Breeding Center for 43 years. After about 87 years, Diego was home. He was at least 110 years old.

Sometimes, humans don't recognize a problem until it's too late and species go extinct. But Española Island is returning to what it looked like 350 years ago before the habitat was almost destroyed by humans.

Diego will live out his years with his children, grandchildren, and great-grandchildren on the island of his birth.

For over 60 years, day after day, year after year, aided by the tortoises themselves, scientists and park rangers hoped and worked and found ways to save the Española giant tortoise from extinction.

Sometimes, humans get it right.

FACTS ABOUT DIEGO

Española Galápagos giant tortoise, *Chelonoidis hoodensis*

SIZE: In 2000, Diego weighed about 198 pounds (90kg).

LIFESPAN: Española giant tortoises can live for well over 100 years old, maybe up to 200 years.

RANGE: These tortoises are native to Española Island and only live there unless taken elsewhere by humans. Currently, a second population lives on Santa Fe Island. This population ensures the survival of the Española species because if anything happens on one island, the other island's tortoises are likely to survive.

LIFE CYCLE: Female tortoises dig a nest about 12 to 15 inches (30 to 40 cm) deep to lay eggs. Depending on the weather and temperature, it may take four to six months for eggs to hatch. The sex of the tortoises is determined by the internal temperature of the nest. With higher temperatures, females hatch. With lower temperatures, males hatch.

TORTOISE OR TURTLE?

Tortoises are a specific kind of turtle. Turtles are reptiles that have a shell, and live in water, on land, or both. A turtle might have webbed feet, flippers, or columnar feet. Turtles can be herbivores or omnivores. Tortoises are turtles that live entirely on land, have columnar feet, and are almost always herbivores.

The Galápagos giant tortoises are adapted to live in arid, desertlike conditions. During the dry seasons on the islands, they can go without food or water.

Originally, there were 15 species of Galápagos tortoise, but three are extinct. How many are left?

LONESOME GEORGE

While the Española tortoise population has recovered, there's another tragic story. On Pinta Island, no giant tortoises had been seen since 1906. Then, in 1971, scientists found one last male, who was named Lonesome George. For years, scientists searched worldwide for a female Pinta tortoise to start a breeding program. While Lonesome George lived at the Tortoise Breeding Center, an information panel near his enclosure warned, "Whatever happens to this animal, let him always remind us that the fate of all living things is in human hands." Sadly, Lonesome George died in 2012, alone, the last of his species. The Pinta giant tortoise is extinct.

CONSERVATION ORGANIZATIONS

The Galápagos National Park was created in 1959 by the Republic of Ecuador to protect the unique habitat and species of the Galápagos Islands. With the establishment of the park, 97% of the islands' land was protected and hunting was prohibited. The Galápagos National Park Directorate manages the park.

The Charles Darwin Foundation was established in1959 and built a research station on Santa Cruz Island. The Charles Darwin Foundation and the Galápagos National Park Directorate have collaborated ever since to study and protect the plants and animals of Galápagos.

Galápagos Conservancy, a nonprofit organization in the United States was established in 1985 to support conservation efforts in the archipelago. The Giant Tortoise Restoration Initiative was launched in 2014 as a joint effort between the Galápagos National Park Directorate and the Galápagos Conservancy to build on the previous 50 years of tortoise conservation. Its long-term goal is to restore tortoise populations to their historical distribution and numbers across Galápagos, including on islands where tortoises went extinct.

GOATS: INVASIVE SPECIES

When sailors first came to Española Island, they wanted tortoises to eat. But soon there were fewer and fewer tortoises. Instead, sailors left goats on the island as an alternative food supply for later voyages. When they returned, the goats had spread across the island, and the sailors had goats to eat.

But goats were an invasive species for these islands. An invasive species is a plant or animal that doesn't live and grow naturally in a place.

The goat's eating habits were different from the giant tortoises. Goats destroyed some plants and let others live. Soon, the island's habitat was very different. For the tortoises to return, the goats had to be removed.

SOURCES

Cayot, Linda J. (Head of Herpetology, Charles Darwin Foundation, 1988-1998; science advisor and later Director of Giant Tortoise Restoration Initiative, Galápagos Conservancy, 2006-2018.) Interview by the author, February 11, 2021.

Gibbs, James P., Linda J. Cayot and Washington Tapia Aguilera, eds. Galapagos Giant Tortoises. London, UK: Academic Press, Elsevier © 2021.

Pinta

ECUADOR

PACIFIC OCEAN

600 miles
(965 km)

Marchena

Genovesa

Santiago

Bartolomé

North
Seymour

Rábida

Baltra

Fernandina

Pinzón

San Cristóbal

Santa Cruz

Santa Fe

10 miles
(16 km)

Isabela

Floreana

Española

THE GALÁPAGOS ISLANDS

The Galápagos Archipelago lies 600 miles (965 km) off the coast of Ecuador in South America. An archipelago is a collection of islands. These were created by volcanic activity and still have 13 active volcanoes. The soils are red, black, brown, and gray. Located near the equator, these tropical islands have a wet season and a dry season.

Española Island, on the southeast corner of the archipelago, is about 8.9 miles (14.3 km) long by 4.25 miles (6.84 km) wide. That's about two-thirds the size of Manhattan Island. It's low, flat, rocky and arid. Scientists estimate that the island once supported maybe as many as 8,000 giant tortoises. It's also home to the world's only significant colony of waved albatross.

The Galápagos Archipelago was brought to world-wide fame when British naturalist Charles Darwin visited there in September 1835 on the *HMS Beagle*. Darwin studied the unique species in the Galápagos, and partly based on his observations there, he wrote his famous book, *On the Origin of Species*. He argued for the theory of natural selection, which explains how species evolve. Since 1960, the Charles Darwin Foundation has maintained a research station on Santa Cruz Island, providing a place for Ecuadorian and international scientists and students to conduct research and to do environmental education for conservation.